Dinosaur HeRAWRsies

Copyright Greystone Studios 2015
By Chandra Reyer and Jennifer Nolan

Illustrations by Chandra Reyer
Cover and interior design by Jennifer Nolan

Dedicated to our dinosaur-loving children, Gillian, Zachary, Julianna & Daniel.

Triceratops
(Three-horned Face)

Triceratops

These little triceratops' frills may have grown to be very colorful.

Kentrosaurus
(Pointed-tail Saurian)

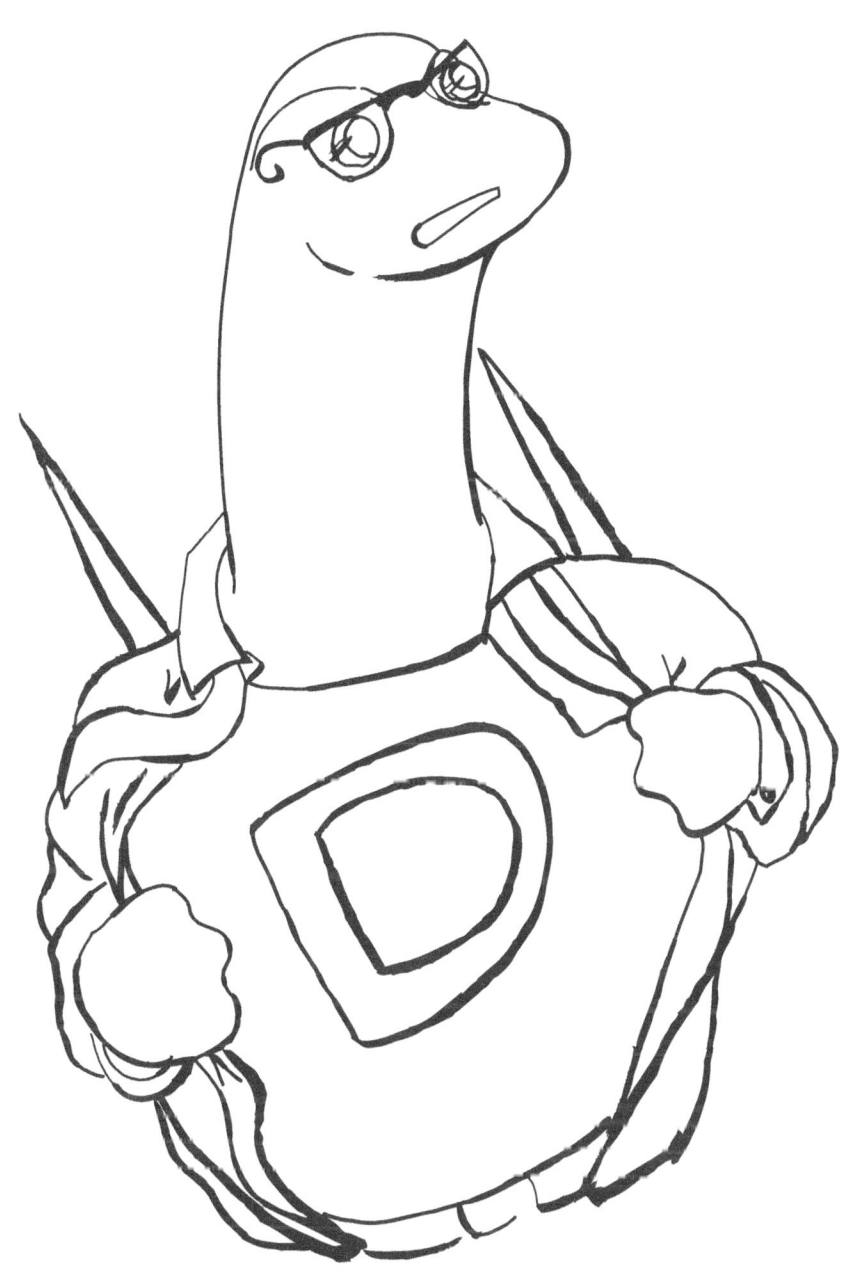

This stegosaur is from Tanzania, not Krypton.

Amargasaurus
(La Amarga Lizard)

The tail and neck spines were taller than any other sauropods.

Tyrannosaurus Rex

(Tyrant Lizard King)

It's possible that T. Rex was actually born with down, much like baby chicks.

Hipster Rex

Eating people is so mainstream...

Sauroniops
(Eye of Sauron)

May or may not have lived on Mt. Doom.

Carnotaurus
(Meat-eating Bull)

Carnotaur's forelimbs were even more useless than T. Rex's!

Tarchia
(Brainy One)

There is no evidence that dinosaurs had books.

Zaraapelta

(Hedgehog, Small Shield)

Zaraapelta was an ankylosaur and did not actually act like a hedgehog.

Humanologists

It's an always-evolving science.

Ouranosaurus
(Brave Lizard)

An interesting African dinosaur with an unusual spine sail related to Iguanodon.

Deinonychus

(Terrible Claw)
(Shown with felis catus... also a terrible claw)

Cats and raptors did not live at the same time.

Helicoprion
(Spiral Saw)

This is not a dinosaur, but a paleoshark with a very puzzling jaw structure.

Aquilops

(Eagle Face)

A recently discovered cat-sized early ceratopsian.

Prestosuchus
(Named after Vicentino Prestes de Almeida, a Brazilian Paleontologist.)

Not actually a dinosaur, but a crocodilian. Crocodiles and birds are cousins.

Suchomimus
(Crocodile Mimic)

Suchomimus is a spinosaur that didn't have a sail on its spine.

Crichtonsaurus
(Named after Michael Crichton, author of Jurassic Park.)

There is no evidence that dinosaurs wrote novels.

Want more dinosaurs?
What about...
Dinosaurs in Space!

Visit Greystone-studios.com to see our other titles including Cosmosaurs: Search for the Hovercat!
For even more dinosaurs, visit idrawdinos.com

Scan QR Code for special offers!

www.ingramcontent.com/pod-product-compliance
Lightning Source LLC
Chambersburg PA
CBHW080621180526
45168CB00007B/3001